SOLAR ECLIPSE ACTIVITY BOOK

DEAR CUSTOMER,

WE ARE DELIGHTED TO WELCOME YOU TO THE WORLD OF WONDER AND DISCOVERY WITH OUR "SOLAR ECLIPSE 2024 ACTIVITY BOOK FOR KIDS." YOUR INTEREST IN OUR BOOK BRINGS A SMILE TO OUR FACES, AND WE ARE THRILLED TO HAVE YOU AND YOUR YOUNG EXPLORERS AS PART OF OUR COSMIC ADVENTURE.

WITHIN THE PAGES OF THIS ACTIVITY BOOK, YOUR CHILDREN WILL EMBARK ON AN EXCITING JOURNEY THROUGH THE UNIVERSE, LEARNING ABOUT THE INCREDIBLE PHENOMENON OF A SOLAR ECLIPSE. PACKED WITH ENGAGING PUZZLES, CREATIVE ACTIVITIES, AND FASCINATING FACTS, THIS BOOK IS DESIGNED TO SPARK CURIOSITY AND IMAGINATION IN YOUNG MINDS.

WE POURED OUR PASSION INTO CREATING THIS ACTIVITY BOOK, WITH THE AIM OF MAKING LEARNING ABOUT THE SOLAR ECLIPSE A THRILLING AND EDUCATIONAL EXPERIENCE FOR YOUR CHILDREN. WE SINCERELY HOPE THAT IT BRINGS THEM COUNTLESS HOURS OF JOY AND ENLIGHTENMENT.

YOUR FEEDBACK AND SUPPORT MEAN THE WORLD TO US. IF YOUR KIDS ENJOY THIS ACTIVITY BOOK AND FIND IT BOTH ENTERTAINING AND EDUCATIONAL, WE KINDLY ASK YOU TO CONSIDER SHARING YOUR THOUGHTS WITH US THROUGH A REVIEW. YOUR VALUABLE FEEDBACK NOT ONLY HELPS US IMPROVE BUT ALSO ASSISTS OTHER PARENTS IN FINDING ENGAGING AND EDUCATIONAL RESOURCES FOR THEIR CHILDREN.

THANK YOU FOR CHOOSING OUR "SOLAR ECLIPSE 2024 ACTIVITY BOOK FOR KIDS." WE LOOK FORWARD TO BEING A PART OF YOUR CHILDREN'S JOURNEY OF DISCOVERY AND HOPE THAT EVERY PAGE OF THIS BOOK SPARKS THEIR CURIOSITY AND BRINGS THEM CLOSER TO THE WONDERS OF THE UNIVERSE.

HAPPY EXPLORING, YOUNG ASTRONOMERS!

SINCERELY,

Zara Harper

WHAT IS A SOLAR ECLIPSE

IN THE MILKY WAY GALAXY, ALL OF THE MOONS AND PLANETS ORBIT AROUND THE SUN. WITH THE CONSTANT MOVING OF THESE OBJECTS, SOMETIMES ONE OBJECT WILL BLOCK ANOTHER OBJECT.
IT IS IMPOSSIBLE FOR OBSERVERS ON EARTH TO SEE THE BLOCKED OBJECT.
THIS IS CALLED AN ECLIPSE

People, especially children, are very curious during a solar eclipse and want to look at it. Looking directly at the solar eclipse may seem safe because you cannot see the sun. However, the harmful rays still exist and should not be looked at directly.

How many smaller words can you make out of the letters from the word:

TELESCOPE

3 LETTERS	4 LETTERS	5 LETTERS	6 LETTERS
1 point	2 points	3 points	4 points

TOTAL POINTS: _____

WITHOUT READING THE STORY, ASK A FRIEND OR FAMILY MEMBER TO HELP YOU FILL IN THE BLANKS WITH DIFFERENT WORDS (LIKE NOUNS, ADJECTIVES, AND VERBS). MAKE SURE EVERY BLANK IS FILLED! ONCE YOU'VE FILLED THEM ALL IN, READ YOUR WACKY STORY ALOUD AND GET READY TO GIGGLE. AFTER THAT, EXPRESS YOUR CREATIVITY BY DRAWING A PICTURE THAT GOES WITH YOUR HILARIOUS TALE. GET READY FOR LAUGHTER AND IMAGINATION!

"HA HA HA"

In a distant land called _____ [place], a rare solar eclipse was _____ [verb] upon them. Excitement filled the air as _____ [adjective] astronomers and _____ [adjective] scientists prepared their _____ [noun] telescopes.

On eclipse day, the sky was _____ [adverb] clear, and the sun shone _____ [adjective] bright. People wore _____ [color] sunglasses and gathered in the square, creating a _____ [adjective] carnival atmosphere.

As the moon slowly _____ [verb] the sun, the world grew _____ [adjective] darker. The sun turned into a _____ [color] ball of fire with a _____ [adjective] corona.

Children were _____ [emotion] excited, and adults marveled at the _____ [adjective] beauty. Some even saw _____ [noun]-shaped shadows.

After _____ [number] minutes, the moon _____ [verb] away, and the world returned to _____ [adjective] normalcy. Cheers erupted, and the eclipse became a cherished _____ [noun] in _____ [place].

It reminded them that in _____ [adjective] lives, there are moments of _____ [adjective] wonder. They knew the next _____ [adjective] eclipse would bring more _____ [adjective] awe.

CONNECT THE DOTS

The two main types of eclipses are solar eclipses and lunar eclipses.

SOLAR ECLIPSE

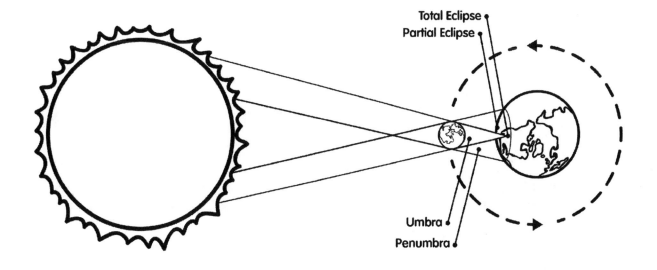

LUNAR ECLIPSE

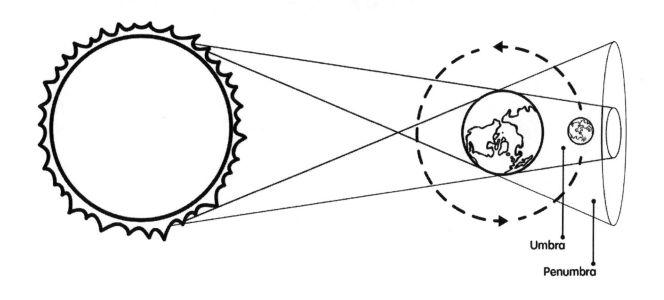

A solar eclipse is when the moon passes in front of the sun. The moon is blocking the sun and therefore casts a shadow in its place. This cannot be seen from everywhere on Earth, only from where the sun's shadow would fall

A lunar eclipse is when the moon passes through the shadow of the Earth from the sun.
There are also three types of lunar eclipses; partial, annular, and total. These
can be viewed without special equipment because the sun is in the opposite direction.

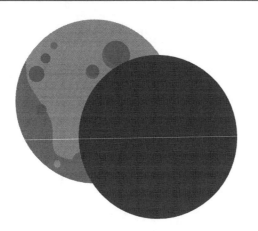

THERE ARE THREE DIFFERENT TYPES OF ECLIPSES.

A partial eclipse is when the moon blocks only a part of the sun.

An annular eclipse is when the moon covers almost all of the sun. The sun can still be seen around the edges of the moon.

A total eclipse would block out the entire sun making it dark for that particular location on Earth.

SOLAR AND LUNAR ECLIPSES

Draw a picture of the position of the moon, sun and earth during a solar eclipse

Draw a picture of the position of the moon, sun and earth during a lunar eclipse

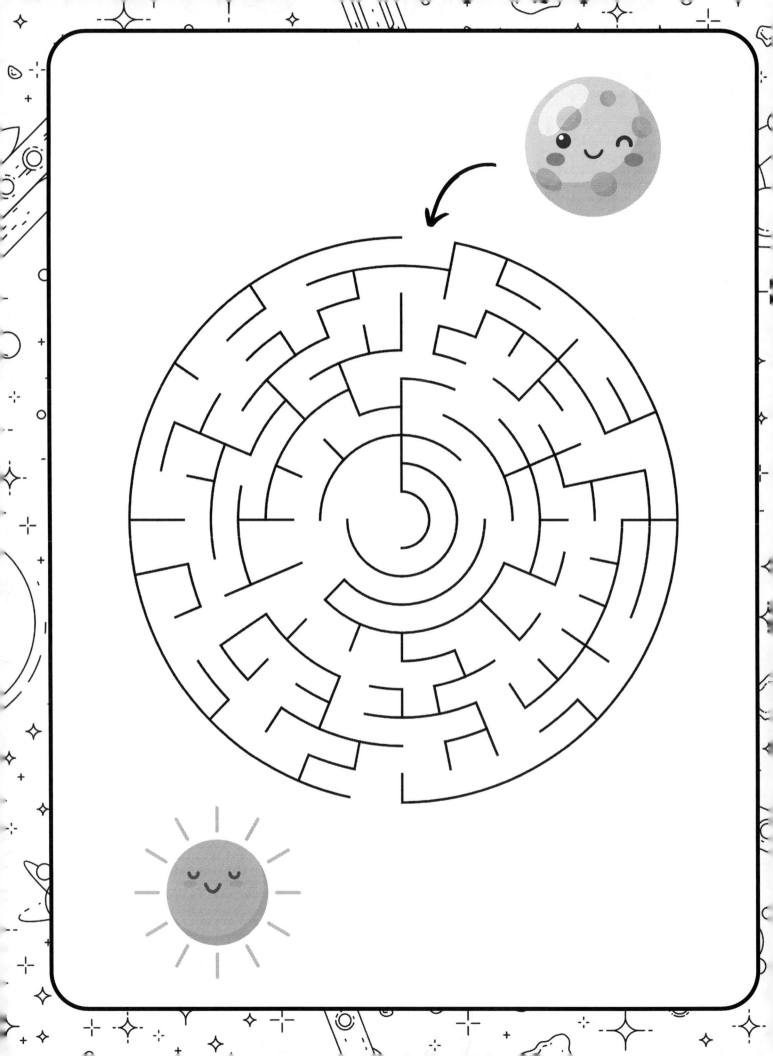

How many smaller words can you make out of the letters from the word:

ECLIPSE

3 LETTERS	4 LETTERS	5 LETTERS	6 LETTERS
___	___	___	___
___	___	___	___
___	___	___	___
___	___	___	___
___	___	___	___
___	___	___	___
___	___	___	___
___	___	___	___
1 point	2 points	3 points	4 points

TOTAL POINTS: ___

Title: "The Great Solar Eclipse Adventure"

Once upon a time in _____ [place], a group of adventurous friends decided to embark on a quest to witness the rare and exciting _____ [adjective] solar eclipse. They gathered their _____ [noun] and set out for the perfect viewing spot. As they reached their chosen _____ [noun], they were greeted by the dazzling _____ [adjective] sunshine. Everyone donned their stylish _____ [noun] and eagerly waited for the eclipse to begin. The excitement in the air was _____ [adjective], and the anticipation was _____ [adjective]. As the moon slowly moved in front of the sun, the sky started to _____ [verb] in a breathtaking array of colors. The birds, not knowing what was happening, began to _____ [verb] loudly, adding to the _____ [adjective] atmosphere. It was a moment of pure _____ [noun] and wonder. The group watched in _____ [adjective] amazement as the moon covered the sun, creating a stunning _____ [noun] in the sky. It felt like they were in a _____ [noun] from another world. The temperature dropped, and a sense of _____ [emotion] overcame them all.

With their _____ [adjective] eclipse glasses, they observed the _____ [noun] around the sun, known as the _____ [noun], and it was more magnificent than they could have ever _____ [verb]. It was a truly _____ [adjective] experience. After the eclipse ended, they packed up their _____ [noun] and headed home, their hearts filled with _____ [noun]. They knew they had shared a _____ [adjective] adventure that they would _____ [verb] for a lifetime.

And so, their Great Solar Eclipse Adventure became a cherished _____ [noun] in their lives, reminding them of the beauty and _____ [noun] of the universe.

```
E V J O M O M E N T F L R D L
T X G T E R G U D E R R N Y E
M H C R P O R M U L S O U A G
Z N R I X H M V R E T G Y K D
E T H P T L Y O A S Y Z Z K T
G C R W D E H B T C G X V F D
Z N L M E C M S I O Y J C B D
K S C I E N C E O P V U Q D K
B A G L P I H R N E I F O W C
B F H Z Q S A V L T K Z P Q O
S E O B D L E E A B Q R B X U
C T M O O N P A O Z J F I W E
L Y W S J D L U P W I B A C F
T O T A L I T Y Z Y X D V E P
X X A N V J U U H Y C W X V L
```

TELESCOPE **EXCITEMENT** **SCIENCE**
MOMENT **SOLAR** **MOON**
TOTALITY **DURATION** **SAFETY**
TRIP **OBSERVE** **ECLIPSE**

SOLAR ECLIPSE CHALLENGE

ACROSS

2. The coolest part of a total eclipse.
4. The planet where we watch eclipses.
7. Days until the eclipse starts.

DOWN

1. Always be safe when watching an eclipse.
3. The path the moon takes around Earth.
5. Eclipses can be a lesson in science.
6. Where you see the eclipse from.

Draw a big circle in the center of the page. This circle represents the Sun.
Now, draw a smaller circle partially covering the Sun. This smaller circle represents the Moon blocking part of the Sun during a solar eclipse.
You can add some lines around the Sun to show rays of light. That's it! You've just created a simple representation of a solar eclipse.
You can now color it in and make it as colorful as you like!

IMAGINE DRAWING A LINE ON A MAP FROM MEXICO TO CANADA, PASSING RIGHT THROUGH THE MIDDLE OF THE UNITED STATES. THIS IS THE PATH OF THE 2024 TOTAL SOLAR ECLIPSE. IT WILL START IN MEXICO AND MAKE ITS WAY THROUGH THESE U.S. STATES: TEXAS, OKLAHOMA, ARKANSAS, MISSOURI, ILLINOIS, KENTUCKY, INDIANA, OHIO, PENNSYLVANIA, NEW YORK, VERMONT, NEW HAMPSHIRE, AND FINALLY, MAINE AND CANADA.

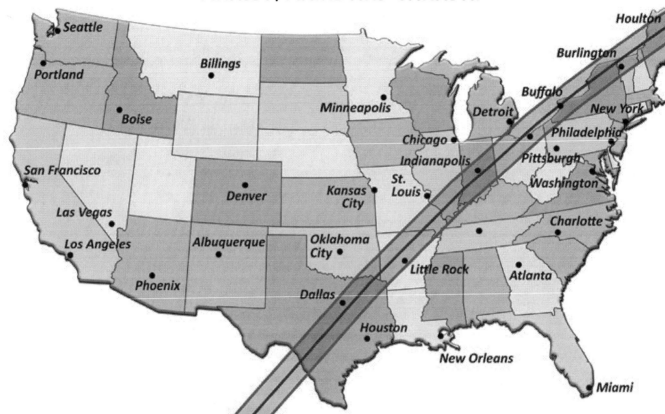

IF YOU'RE LUCKY ENOUGH TO BE IN THE PATH OF THE TOTAL ECLIPSE, YOU'LL EXPERIENCE SOME AMAZING THINGS:

Totality: This is the special moment when the Moon completely covers the Sun, and everything goes dark. You'll see the Sun's outer atmosphere called the corona, which looks like a beautiful, glowing halo. It's safe to look at the Sun ONLY during totality.

Partial Eclipse: If you're not in the path of totality, you'll still see a partial eclipse. It'll look like a bite taken out of the Sun. Remember to use special eclipse glasses to protect your eyes while watching the partial eclipse.

Excitement: No matter where you are, there will be an air of excitement. People from all over the world will come to see this incredible event, and you'll have a chance to meet new friends and learn from others.

Rearrange the mixed-up letters to find the hidden words and enjoy the game!

ATRHE _ _ _ _ _

EVSNU _ _ _ _ _

SARNUU _ _ _ _ _ _

MSAR _ _ _ _

ONOM _ _ _ _

NUS _ _ _

JITEPRU _ _ _ _ _ _ _

EYRURCM _ _ _ _ _ _ _

NNEPUTE _ _ _ _ _ _ _

SRTAUN _ _ _ _ _ _

TUPLO _ _ _ _ _

ARST _ _ _ _

SNOMYORTA _ _ _ _ _ _ _ _ _

BTLES _ _ _ _ _

CMTOE _ _ _ _ _

```
K C J W P U L I C T U O J D
T T E D U C A T I O N O H A
L Z V M Q H D W H S Q K Y R
C H R M P A R T I A L H V K
A O G Y E K S I D Q T S U N
B R R X B V P S H A D O W E
K I W R T T E O P W R O G S
N Z C S H W C N J E D J L S
T O T A L I T Y T T P T A O
K N O R U D A W N G S T S U
L W Y Z X O T U L R T H S V
Y H E P R G O H X T L U E K
Y C O H R C R J I P H O S X
Q T N O W T O Y D T J F S Z
```

SPECTATOR   **EDUCATION**   **COUNTDOWN**
SUN   **GLASSES**   **SHADOW**
PATH   **HORIZON**   **DARKNESS**
EVENT   **PARTIAL**   **TOTALITY**

CONNECT THE DOTS

SUN AND MOON PUZZLE FUN

ACROSS

5. What happens during totality.
7. Eclipses can happen during this.

DOWN

1. Learning about eclipses is fun!
2. The route where the eclipse can be seen.
3. When only part of the sun is covered.
4. Eclipses make us marvel at the sky.
6. Another word for the sun.

TRUE OR FALSE

A SOLAR ECLIPSE OCCURS WHEN THE MOON PASSES BETWEEN THE EARTH AND THE SUN.

TRUE ✔ ✘ FALSE

A SOLAR ECLIPSE CAN BE SEEN FROM ANYWHERE ON EARTH.

TRUE ✔ ✘ FALSE

DURING A TOTAL SOLAR ECLIPSE, IT BECOMES COMPLETELY DARK ON EARTH, SIMILAR TO NIGHTTIME.

TRUE ✔ ✘ FALSE

IT IS SAFE TO LOOK AT A SOLAR ECLIPSE WITH THE NAKED EYE AS LONG AS IT'S NOT FOR AN EXTENDED PERIOD.

TRUE ✔ ✘ FALSE

DURING AN ANNULAR ECLIPSE, THE MOON COVERS ALMOST ALL OF THE SUN, BUT THE SUN CAN STILL BE SEEN AROUND THE EDGES OF THE MOON.

TRUE ✔ ✘ FALSE

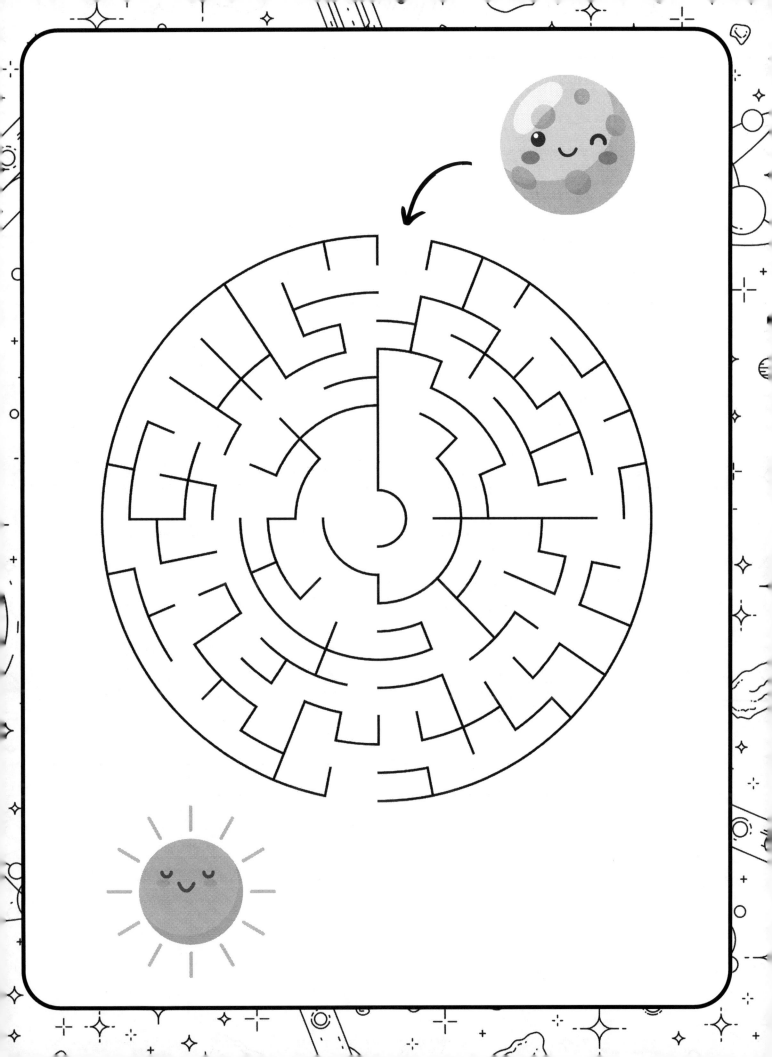

How many smaller words can you make out of the letters from the word:

ASTRONOMY

3 LETTERS	4 LETTERS	5 LETTERS	6 LETTERS
_____	_____	_____	_____
_____	_____	_____	_____
_____	_____	_____	_____
_____	_____	_____	_____
_____	_____	_____	_____
_____	_____	_____	_____
_____	_____	_____	_____
_____	_____	_____	_____
1 point	2 points	3 points	4 points

TOTAL POINTS: _____

In _____ [place], young Timmy's _____ [adjective] excitement for the upcoming solar eclipse knew no bounds. He simply couldn't wait for the big day.

With a _____ [noun] on his head and a _____ [noun] in hand, Timmy believed he could make the eclipse happen _____ [adverb]. He chanted, "Eclipse, eclipse, come to me, I want to see you now, you see!"

Neighbor Mrs. Jenkins chuckled and called, "Timmy, you can't rush a solar eclipse! It has its own _____ [noun] !"

Undeterred, Timmy wore his pajamas _____ [adverb] and danced a _____ [adjective] jig.

Even his dog, _____ [dog name], joined in the fun.

Days passed, and Timmy tried all sorts of _____ [adjective] tricks, even offering his favorite _____ [snack] as a bribe.

Finally, on eclipse day, Timmy lay on the grass, gazing at the sky. As predicted, the moon began to slowly cover the sun. Timmy realized that the universe had a _____ [adjective] sense of humor.

With totality, Timmy cheered. He learned that some things were worth _____ [verb ending in -ing] for, and that the universe could make even the _____ [adjective] dreams come true.

In the midst of the eclipse, Timmy, pajamas on backward, shared this lesson with laughter and the unexpected.

CONNECT THE DOTS

```
O S D N W O F L U F P L L Q N S B
A W C C I N D I A N A E G S W R I
T N E W Y O R K L R E U I A F U H
Q E U P M W P G W S K V D D O M N
X P X P M Z K S U J E A G Y S K A
U T E A T J I M Z U K N U S S M
C P W N S O V R T U T G L S R H G
P E V U N W U O G N U B I Q A S Q
B U A I M S F W O X C H O K S J
O P L M J B Y M M N K N K H T N Z
H L V M Q B R L L I Y M L C N A N
I S U U K E M W V Y S K A Y T Q N
O M K D V G F U G A G S H I A B G
B T C F C O D D N N N S O U N C C
W K M L M B Z G G M G I M U T E P
A U S M O A H B B A X T A A R V P
Z H L T C H S U B T J D Q W S I J
```

TEXAS   OKLAHOMA   ARKANSAS

MISSOURI   ILLINOIS   KENTUCKY

INDIANA   OHIO   PENNSYLVANIA

NEW YORK   VERMONT   MAINE

ECLIPSE EXPLORERS CROSSWORD

ACROSS

3. When the sun is completely covered.
4. It gets covered during an eclipse.
5. The moon's makes one during an eclipse.
7. When the sun, moon, and Earth line up.

DOWN

1. Causes the eclipse by blocking the sun.
2. You need these to look at the sun safely.
6. What the sky becomes during an eclipse.

SOLAR ECLIPSE

Name: _____   I am [ ] years old

DID YOU WEAR SOMETHING TO SEE THE SUN?
Yes | No

I will remember most of the eclipse is

" _____
_____
_____ "

THIS IS WHAT A THE ECLIPSE LOOKED LIKE:

The weather was _____   **DID IT GO TOTTALY DARK?**

I watched the eclipse at _____   Yes | No

SOLUTIONS

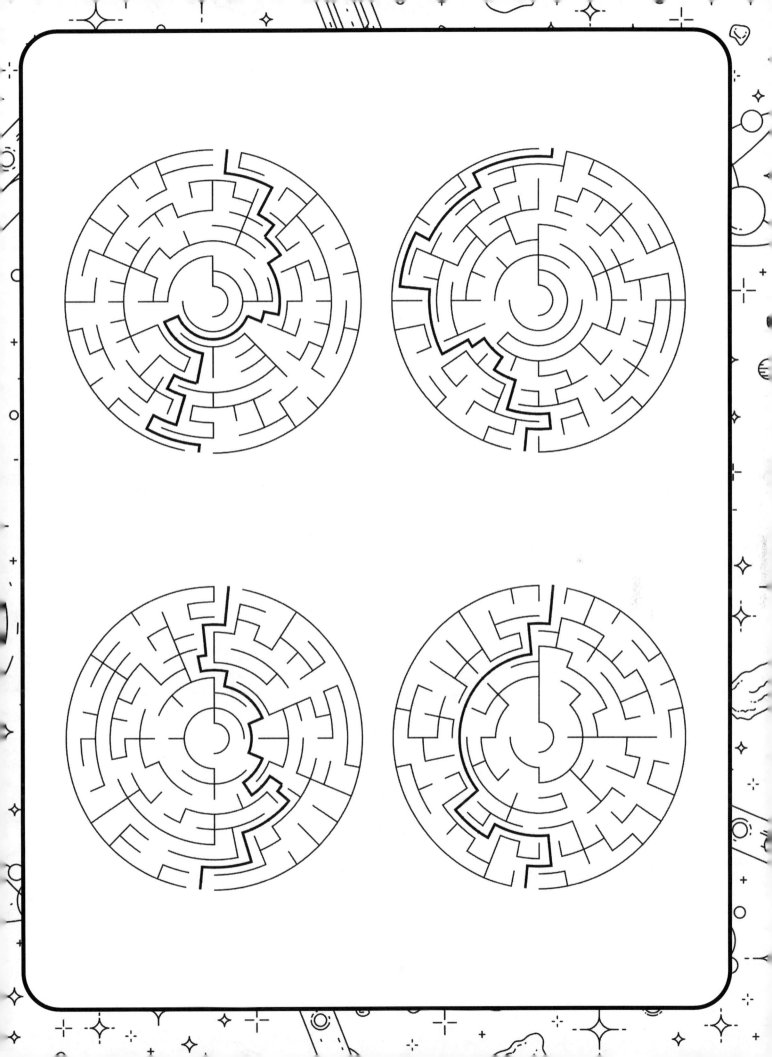

ATRHE	E A R T H
EVSNU	V E N U S
SARNUU	U R A N U S
MSAR	M A R S
ONOM	M O O N
NUS	S U N
JITEPRU	J U P I T E R
EYRURCM	M E R C U R Y
NNEPUTE	N E P T U N E
SRTAUN	S A T U R N
TUPLO	P L U T O
ARST	S T A R
SNOMYORTA	A S T R O N O M Y
BTLES	B E L T S
CMTOE	C O M E T

```
E V J O M O M E N T F L R D L
T X G T E R G U D E R R N Y E
M H C R P O R M U L S O U A G
Z N R I X H M V R E T G Y K D
E T H P T L Y O A S Y X Z K T
G C R W D E H B T C Q Y F D
Z N L M E C M S I O Y J C B D
K S C I E N C E O P U Q D K
B A G L P I H R N E I F O W C
B F H Z Q S A V L T K Z P Q O
S E O B D L E E A B Q R B X U
C T M O O N P A O Z J F I W E
L Y W S J D L U P W I B A C F
T O T A L I T Y Z Y X D V E P
X X A N V J U U H Y C W X V L
```

```
K C J W P U L I C T U O J D
T T E D U C A T I O N O H A
L Z V M Q H D W H S Q K Y R
C H R M P A R T I A L H V K
A O G Y E K S I D Q T S U N
B R R X B V P S H A D O W E
K I W R T T E O P W R O G S
N Z C S H W C N J E D J L
T O T A L I T Y T T P T A O
K N O R U D A W N G S T S U
L W Y Z X O T U L R T H S V
Y H E P R G O H X T L U E K
Y C O H R C R J I P H O S X
Q T N O W T O Y D T J F S Z
```

```
O S D N W O F L U F P L L Q N S B
A W C C I N D I A N A E G S W R I
T N E W Y O R K L R E U I A F U H
Q E U P M W P G W S K V D D O M N
X P X P M Z K S U J E A G Y S K A
U T E A T J I M Z U N K N U S S M
C P W N S O V R T U T G L S R H G
P E V U N W U O G N U B I Q A S Q
B U A I M S F W O X C H O K V S J
O P L M J B Y M M N K N K H T N Z
H L V M Q B R L L I Y M L C N A N
I S U U K E M W V Y S K A Y T Q N
O M K D V G F U G A G S H I A B G
B T C F C O D D N N N S O U N C C
W K M L M B Z G G M G I M U T E P
A U S M O A H B B A X T A A R V P
Z H L T C H S U B T J D Q W S I J
```

SOLAR ECLIPSE CHALLENGE

ACROSS

2. Totality - The coolest part of a total eclipse.
4. Earth - The planet where we watch eclipses.
7. Countdown - Days until the eclipse starts.

DOWN

1. Safe - Always be safe when watching an eclipse.
3. Orbit - The path the moon takes around Earth.
5. Teach - Eclipses can be a lesson in science.
6. Horizon - Where you see the eclipse from.

SUN AND MOON PUZZLE FUN

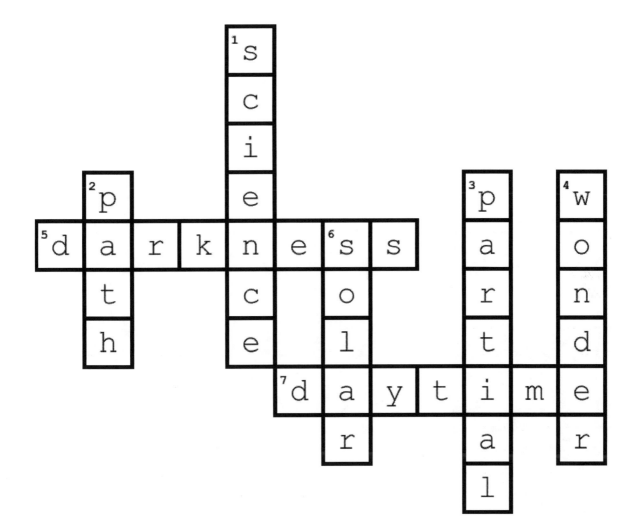

ACROSS

5. Darkness - What happens during totality.
7. Daytime - Eclipses can happen during this.

DOWN

1. Science - Learning about eclipses is fun!
2. Path - The route where the eclipse can be seen.
3. Partial - When only part of the sun is covered.
4. Wonder - Eclipses make us marvel at the sky.
6. Solar - Another word for the sun.

ECLIPSE EXPLORERS CROSSWORD

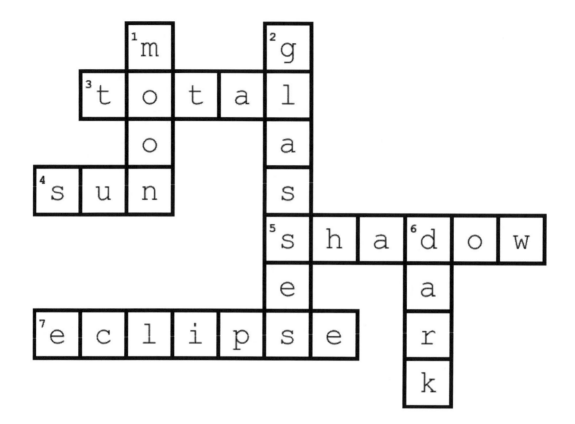

ACROSS

3. Total - When the sun is completely covered.
4. Sun - It gets covered during an eclipse.
5. Shadow - The moon's makes one during an eclipse.
7. Eclipse - When the sun, moon, and Earth line up.

DOWN

1. Moon - Causes the eclipse by blocking the sun.
2. Glasses - You need these to look at the sun safely.
6. Dark - What the sky becomes during an eclipse.

TRUE OR FALSE

A SOLAR ECLIPSE OCCURS WHEN THE MOON PASSES BETWEEN THE EARTH AND THE SUN.

TRUE

A SOLAR ECLIPSE CAN BE SEEN FROM ANYWHERE ON EARTH.

FALSE

DURING A TOTAL SOLAR ECLIPSE, IT BECOMES COMPLETELY DARK ON EARTH, SIMILAR TO NIGHTTIME.

TRUE

IT IS SAFE TO LOOK AT A SOLAR ECLIPSE WITH THE NAKED EYE AS LONG AS IT'S NOT FOR AN EXTENDED PERIOD.

FALSE

DURING AN ANNULAR ECLIPSE, THE MOON COVERS ALMOST ALL OF THE SUN, BUT THE SUN CAN STILL BE SEEN AROUND THE EDGES OF THE MOON.

TRUE

Made in United States
Orlando, FL
14 March 2024